SACRAMENTO CITY

Suttersville

Swartz Ranch

Webster

SACRAMENTO

TO

Tule

Marsh

Swartz Slough

Merritt Slough

Georgina Channel

WITHDRAWN

WORN, SOILED, OBSOLETE

Mòquelumne River

Laird's

Lower Ba

ylor

addon's

Hick's

Slough

New York

SAN JOAQUIN

Davis & A

Calaveras River

Isbel's

Good

Stockton Slough

STOCKTON

Tule

French Creek

Castoria

Heath

Islip's

San Joaquin City

Stanislaus City

Colton's

D0792310

Eureka! I have found it! One thinks immediately of the struggling miner, far from home, challenged by the elements and a new land, competing with others for dwindling riches, and the exclamation made upon the reward of hard work or a chance find. One thinks of the California gold rush.

That singular triumph of finding a life-altering treasure is but a tiny facet of a global story: the world rushing to California in 1848 through the early 1850s. The story is much, much deeper. There is the devastation to California's Indian population and the dislocation of the ranchos and Spanish land grants. There are the stories of the Chinese, French, Sonorans, Hawaiians, Australians, and others from around the globe who joined the adventure. There are the stories of African Americans who secured their freedom, and of women, and of entrepreneurs who made more money supplying the miners than the miners made through their own labors.

The range of expression by participants is wide. On hearing of the discovery of gold, James Carson wrote, "A frenzy seized my soul . . . piles of gold rose up before me . . . castles of marble . . . thousands of slaves . . . myriads of fair virgins . . . were among the fancies of my fevered imagination. In short, I had a very violent attack of the gold fever."

It took but a short time for the realities of arduous labor to set in. William Swain, one of the intrepid argonauts, suffered tremendous hardship, first traveling across the country by wagon. And then, after a few months, he writes, "Almost every man we meet who has been in the mines is wishing himself out of this country. . . . Generally speaking, the gold fever cools down after a man has been here a week or two."

There followed shortly environmental degradation on a grand scale and impact that was hard fought (and won) by agricultural enterprises downstream from hydraulic mining. There also quickly followed the risk-taking businessmen such as Collis Huntington, Mark Hopkins, Charles Crocker, Leland Stanford, Levi Strauss—the builders of railroads, cities, and manufacturing empires.

More significantly, there came California—the political entity, the landscape, and the state of mind we have today. Then came the California Dream—an extension of the American Dream, the pursuit of happiness.

This small book captures the spirit of the California gold rush and touches on its consequences. Those who see today's Californians embracing freedom of dress, of purpose, and of thought will perhaps understand that the roots of that expression go back to those days when the world rushed in with high hopes and boundless energy. In computer development, information technology, biomedical engineering, graphic design, the wine industry, food, and art we see that spirit in the many new gold rushes of today.

Dennis M. Power
Executive Director
Oakland Museum of California

GOLD FEVER

CALIFORNIA'S GOLD RUSH

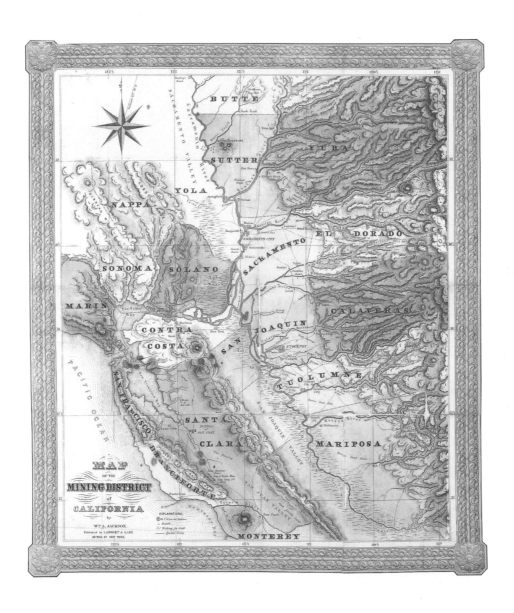

A CLOSE-UP GUIDE

CALIFORNIA IS A PLACE OF GOLDEN SUNSHINE,
BIG MOUNTAINS, TOWERING TREES, AND TALL STORIES,
SOME OF THEM TRUE.

It even began as a legend. The first known mention of California, recorded in sixteenth-century Spain, was a report of a fantastic island "very near the terrestrial paradise," ruled by the goddess Califia and inhabited by Amazons. "Their arms are all of gold as is the harness of the wild beasts they ride. In all the island there is no other metal."

California, of course, is no island, and there are no Amazons. But the myth of the golden land turned out to be reality.

American California was born in gold, developed instantly in an amazing mass migration called the gold rush that has characterized the state ever since.

The gold rush of 1849 transformed California almost overnight from a sleepy, pastoral spot at the far end of the world into the thirty-first of the United States.

Today some 33 million people live in California, making it the largest state in the country. It's a land of limitless horizons, the home of Silicon Valley and the movies, of cities big and small. If it were a country, its economy would rank seventh in the world.

It is hard to imagine now what California looked like in the middle of the nineteenth century, before American rule and before the gold rush.

(above) A 1950 postage stamp commemorating the centennial of California statehood depicts a '49er panning for gold.

(opposite page) The ultimate tool of the gold miners—the pan, with a few nuggets of gold in it. This tool was where the prospector placed all his hopes.

(previous pages) (left) An exquisite Chinese chest, c. 1875, with a fanciful view of America, complete with Indians pursuing buffalo across the plains. (right) Map of the original mining districts of California, c. 1851

(above) The landscape of
a golden California:
A.D.O. Browere's South
of Tuolumne City, painted
in 1861.

It had been inhabited by 300,000 Native Americans, divided into hundreds of tribelets. Then Spanish settlers came from Mexico, bringing Christianity, as well as smallpox, measles, and European ideas of how people should live.

By 1841, only half the Indians remained. The rest had died of diseases imported from Europe or from the impact of contact with European civilization, which destroyed their culture and scattered their families. The lords of the land were Mexican, presiding over a few small towns, a handful of crumbling Spanish missions, and vast herds of cattle.

There may, in fact, have been more cattle than people. Only 7,500 people of European descent—fewer than the present population of Sausalito—lived in an area as big as France.

Into this area, in the summer of 1841, came a party of Americans who had traveled more than a thousand miles from the Missouri River across uncharted and nearly unknown country. Their wagons were the first wheeled vehicles to cross the continent.

They arrived at a ranch owned by a Dr. John Marsh in what is now Contra Costa County. "There were no other settlements in the valley," wrote John Bidwell fifty years later. "It was, apparently, still just as new as when Columbus discovered America, and roaming over it were thousands of wild horses, of elk, and of antelope."

Five years later, the United States, riding a wave of what was called Manifest Destiny, declared war on Mexico and took the virtually unpopulated land. It was meant to be a preemptive strike:

(above) A California Indian in the 1850s. When the gold rush began, there were 150,000 native Californians. By 1860 only 36,000 were left.

(opposite) The Spanish came to California for the greater glory of God—and for land. These crucifixes were used by missionaries at Carmel in the 1790s.

(above) John A. Sutter in
the uniform of a general
of the California militia.
A great man before the
discovery of gold, a
pauper afterward.

Surely this fair land could not stay as it was. France or, more likely,
England might take it.

But it became American, and by 1848 the white population of the new
territory had nearly doubled, to about 13,000. Still, there were no cities
and only a few real towns. The biggest and most promising in the northern part of the state was the tiny port of San Francisco.

Except for Sutter's Fort, located a hundred miles up the winding rivers
from San Francisco, the interior of California was a howling wilderness.
There were settlements in only a few places, and one of these was the little Sierra foothill valley the Indians called Culomah, where workmen
overseen by an American named James Marshall were building a sawmill.

One chilly January morning, Marshall, worried that some work had not
been done properly, set out walking to make an inspection. His job was to
ensure that water diverted from the south fork of the American River would
turn the waterwheel and thus drive the mill's machinery. Stopping to look

into the water, Marshall spotted a brilliant, golden rock glittering there. He picked it up, examined it, rolled it between his fingers. "It made my heart thump," he wrote later, "for I was sure it was gold."

Marshall rushed over to the others in his crew. "Gold, boys, gold!" he yelled. "By god I believe I have found a gold mine."

California was born that day, January 24, 1848.

Marshall told his employer, the Swiss pioneer John Augustus Sutter, of his discovery. Sutter told him to keep quiet. But, of course, he could not.

When word got out, it set off a huge rush to California, the largest mass migration in American history, according to historian Michael Kowalewski. People came from all over the world. California's population jumped in six years to 300,000, and the tiny port of San Francisco became the tenth largest city in the United States.

(opposite) James Marshall, who discovered the first gold in January 1848, used this hammer. Marshall, however, never found gold again and died broke.

(above) By 1853 the tents and slapped-together shacks started to be replaced by substantial towns in the Mother Lode as depicted in this view of the main street of Murphys near Angels Camp.

(above) After he lost his Sacramento property, John Sutter moved to his Hock Farm on the Feather River. Within a few years, he had lost even that.

It was an amazing event. The new Californians—calling themselves 49ers—built a hundred mining camps, dug up the country, gouged the hills, and changed the place forever. They also despoiled the land, shot the peaceful Indians living in California, and exterminated the wild game. It was a reckless and dangerous time.

The sudden boom, the big transformation of California, the development of new technology, was to be repeated over and over throughout the state's history. The gold rush, said historian Kevin Starr, "is part of the DNA of California."

The gold miners did not make California, but the discovery of gold led to the growth of service industries. The miners had to be fed, supplied, housed, and transported to the gold fields. Gold lured Levi Strauss, and he sold the miners pants—Levi's they were called. The 49ers drank steam beer, ate sourdough bread, shipped their goods on Wells Fargo express, all staples of modern California. Later came the Pony Express and the pioneer wagon trains and all the myths of the old west, cowboys, Indians, miners, and vigilantes.

Amazingly, not everyone believed the stories of gold at first. Down in San Francisco, that rainy winter, they heard the tales and laughed.

Sutter, who owned the land, or thought he did, wanted to keep things quiet. He swore his men to secrecy and tried to get a land title, which he'd purchased from the local Indians for about $150, recorded in the record books.

But Colonel Richard Mason, the American governor, would have none of it. There would be no land deal with Indians, he said. The American government recognized no Indian claims to California land.

That left Sutter in a difficult situation. Gold had been found on land he was working but did not own. His crews were picking up nuggets on their lunch break and after work. Soon they stopped working for him and went into the gold business. It was like the current-day lottery: why work for wages when you can pick up gold right out of the ground?

Sutter was haunted by the idea of gold and of the thousands who would come to take his land. "He saw the curse on him," wrote historian Hubert Howe Bancroft.

(above) The dream that drew thousands to California: gold nuggets, glittering and unmistakable.

(below) The dream of every argonaut was the gold strike—a creek or river with gold for the taking. This one was the elusive El Dorado in the Sierra foothills.

But all that was in the future, and the word was slow to get out.

"The people did not believe it," said a man named Findla. "They thought it was a hoax. They had found [gold] in various places around San Francisco . . . specimens of different minerals, gold and silver among them, but in very small quantities. And so they were not inclined to believe in the discovery at Sutter's Mill."

(above) The Mexicans, who called themselves Californios, were famous horsemen. This set of silver inlaid spurs was worn by the well-off rancheros, the lords of the land in pre-American times.

Coal had been discovered on Mount Diablo, and, said Robert Semple, "I would give more for a good coal mine than for all the gold mines in the universe."

One of the two newspapers in San Francisco sent E. C. Kemble to the foothills, where he went, as he put it, "to ruralize among the rustics of the country for a few days.

"Great country," he wrote. "Fine climate . . . full flowing streams, mighty timber, large crops, luxuriant clover, fragrant flowers, gold and silver."

Reports of gold mines, he concluded, "a sham."

But into this self-satisfied little town of San Francisco came one Samuel Brannan, an elder of the Mormon Church and a man who knew the main chance when he saw it. He rode up and down Montgomery Street, waving a bottle full of gold and shouting, "Gold! Gold from the American River!"

Brannan touched off a gold rush from the San Francisco area, as he had intended. But first he had taken care to corner a large supply of mining equipment, picks and shovels and pans, which he sold to would-be prospectors at what were then called "patriotic prices." He became California's first millionaire.

(above) San Francisco was a city of shacks, windswept and rude, in the summer of 1849. Telegraph Hill is in the center.

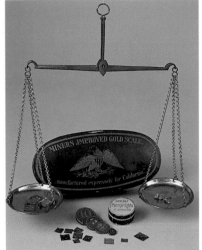

Anything was possible, or so the people thought. Small boats that sold for $50 in the winter now went for $500 and were sailed or rowed up the bay, upriver in the general direction of the gold diggings.

In March, the Martinez-Benicia ferry was making one trip a day. By May there were 200 wagons waiting to cross the Carquinez Strait and head for gold. The boat ran continually day and night.

By June, San Francisco had been abandoned, "Three-quarters of the men," said Bancroft, "had gone to the mines. It was as if an epidemic had swept the little town. . . ." The streets were as empty "as if it was always early morning there."

In Monterey, Sergeant James Carson of the U.S. Army heard the news and "a frenzy seized [his] soul." He had "a very violent attack of gold fever . . . [and] armed with a wash hand basin, fire shovel, and a rifle . . . " he was soon on his way "at high pressure mule speed for the diggings."

"This little scratch upon the earth to make a backwoods mill race touched the cerebral nerve that quickened humanity and sent a thrill through the system," Bancroft wrote. "It tingled the ear and the finger ends; it buzzed about the brain and tickled the stomach, it warmed the blood and swelled the heart."

It was the stuff of dreams. "Castles," Bancroft said, "were built in the air."

(opposite) A lettersheet, sort of an early-day illustrated story, sold to miners to be sent home. This one shows James Marshall and the sawmill where he discovered gold in 1848.

(above) A small scale used for weighing gold. Gold is always heavier than base metal, and the scale measured how much gold the miner might have.

(below) The classic sluice box. Crushed ore went in the top, water was run through it, and the miner watched carefully for the unmistakable glitter of gold.

(above)
*Thousands of Chinese
came to California
during the gold rush.
This man wore the long
queue common among
nineteenth-century sub-
jects of the Chinese
empire.*

There seemed to be gold everywhere in the foothills from the northern Sierra to the south. So much gold was there it could be dug up with spoons, stored in bottles, gambled away, wasted. There was gold for everyone.

By July, there were 4,000 men in the gold country, digging and panning the rivers and creeks. Word spread like ripples from a rock dropped in a pond. Parties came from Los Angeles, and then from Oregon, from Mexico, and from the Sandwich Islands, as Hawaii was then called.

After a golden fall, winter closed in on the mountains of California, and, with the onset of the difficult rainy season, the first year of the gold rush came to an end.

Rumors, however, had finally reached the East. Reports were published in the Baltimore and New York press in the late summer of 1848 only to be regarded with skepticism—another tall tale from the Far West.

But in December, President James K. Polk sent a message to Congress stating that large quantities of gold had been found in California. It was in his interest to say so, since he had presided over the war with Mexico that brought California under American rule. "Mr. Polk's war," they called it, and many had said it was wrong, including a congressman from Illinois named Abraham Lincoln.

Now Polk could say, in effect, that California was full of treasure. The rush was on.

As Bancroft described it: "The trader closed his ledger to depart, and so the toiling farmer . . . the briefless lawyer, the starving student, the quack, the idler, the harlot, the gambler, the henpecked husband, the disgraced, with many earnest, enterprising, honest men and devoted women. Those and others turned their faces westward, resolved to stake all upon a cast; their swift thoughts, like the arrow of Acestes, taking fire as they flew."

The word spread at the turn of the year. This was perfect timing, for the world was in ferment. In 1848 there had been revolutions in France, Germany, Austria, and Hungary. The revolts had been crushed and the rebels were ready to dream dreams of a country where the land was free and there was gold.

In China there was a famine, and thousands of men there had the same dream.

When the steamer *California* sailed from New York for California in the fall of 1848, there were no passengers for San Francisco. But by the time the ship rounded Cape Horn and reached Peru, word of gold was everywhere.

As the ship arrived in Panama, a crowd of gold seekers waited on the beach, willing to pay any price to be taken to California. The skipper allowed more than 300 aboard. Considering there were berths for only 60, the scene must have been chaotic.

And the passengers were hardly saints. Captain Cleveland Forbes thought a good many of them were "the scum of creation, black legs, gamblers, thieves, runners, and drunkards." They started fires, stole from each

(opposite)
When the Chinese came to California, they brought with them the symbols of the land they left behind —the queues, or long pigtails of hair, showed their subjection to the Manchu emperor in China.

(below)
Made in China for the California trade, chests like this were used by the old Spanish and Mexican families to store religious vestments, lace, silk, or other finery.

THE WAY THEY GO TO CALIFORNIA.

(above)
There were two ways to get to California—by sea or by land. But in the imagination of the day, some went by rocket ship or by airship. Real travelers would have to wait a century to do the same thing.

other, and fought. The stokehold crew mutinied, and at least one passenger claimed the officers and seamen were so incompetent as to make the situation "unsafe, dangerous and sometimes critical."

The engineers ran out of fuel, so they had to put in to Monterey and send a party ashore to chop down trees. Later, they burned some of the ship's furniture.

When the *California* steamed into the Golden Gate on February 28, 1849, flags and sirens greeted her. She was the first steamer ever to make California on her own power, and her arrival was the official beginning of the gold rush.

It had been an epic voyage, 18,000 miles around South America. The crew, of course, immediately jumped ship. Only the captain and the third assistant engineer stayed onboard.

Meanwhile in New York, Boston, and every port on the East Coast, hundreds of ships, some seaworthy, some not, made ready to sail to California.

"Preparations were at once commenced to cross the sea, to go to California . . ." the *New York Herald* reported. "There was a general ignorance in the community as to the whereabouts of this new region. Geography has never been a favorite study with the democracy. . . ."

The papers were full of advertisements for new and fast clipper ships, swift steamers, and elegant vessels of every description.

The reality was quite different. You had two choices: Sail to Panama, cross the treacherous isthmus by canoe and mule, then head up the west coast—if it were possible to find a ship. Or, travel around Cape Horn, the worst sea trip in the world. Cape Horn—known as "Cape Stiff" by sailors—was dangerous and often deadly.

The latter route took three months at least, often much longer.

George Dornan, then 19, wrote of first sighting San Francisco Bay from the sailing ship *Panama* in the summer of 1849: "We passed rapidly eastward though the Golden Gate, and as we entered the fog entirely lifted and the sun shone out brightly, giving us a cheering welcome. . . ."

They followed a British ship, with its flags flying and its band playing, into the harbor. "We passed up the channel, rounded Clark's Point, and at about six o'clock on the eighth day of August, 1849, cast anchor opposite the cove on the slopes of which were located the tents and shanties then constituting the infant city of San Francisco."

Dornan remembered his first day in California: "Of California news I can only remember how I drank it in, open mouthed, for all that they said, so much of which sounded like romance; my head whirled and my brain tired in the effort to grasp it all; there was no sleep for me. . . ."

By November of 1849 there were 600 ships anchored at San Francisco and 500 of them had been abandoned. San Francisco was built on the bones of ships that still lie

(above) A percussion rifle and knife from 1850.

(below) Large sluice box operations turned the sites of ancient rivers into ditches and devastated the countryside.

(above) Operating a sluice box on the Mother Lode. Fred Stocking and his wife, Lucinda, at Big Oak Flat, Tuolumne County, 1856.

(opposite) Not everyone in the gold rush sought gold. This advertisement heralded jobs for laborers, lots for sale, and offers to buy gold dust.

beneath the modern-day city. (Construction of a building in the early 1990s was stopped for days so urban archaeologists could dig around the remains of a gold rush ship.)

In the spring of 1849, the population of San Francisco numbered some 2,000 souls, but by December it was ten times that.

The historian Bancroft said that lots offered at $5,000 with no takers were sold the next day for $10,000, and that by the fall of 1849 eggs were $1 apiece—about $17 in today's money. One could buy an omelette in a restaurant for what would now be $34.

"It was a period of fancy prices," he said. A lot that sold in 1847 for $16.50 went at the beginning of 1849 for $6,000 and at the end of the year for $45,000.

Mining was a job for chumps. The real money was to be made in San Francisco, where anyone who could handle a saw or a hammer got $12 a day. And in Sacramento, it was said, a carpenter could make even more—$16 a day, a princely sum.

One prostitute, historians claim, made $50,000 in a single year.

100
LABORERS

Wanted to build a Wharf and erect the Marine Telegraph in this town.   Also, 50 Carpenters, 10 Stone Masons, 20 Pile Drivers, and 20 men for general work.

Proposals will be received up to the 1st April, 1849, for furnishing 1,000 bbls of Shell or Stone lime, 250,000 Bricks, hard burnt, and 250 cubic yards of rubble or broken stone.

T. J. ROACH, Agent for Marine Telegraph,

10,000

Pounds of Gold Dust wanted immediately, for which the highest market price will be given in Gold or Silver Coin.

U. S. TREASURY NOTES,

Drafts on London, Paris, New York, Philadelphia, Boston, and Baltimore.   Also Valparaiso, Mazatlan, Hong Kong China, and Sidney New South Wales.

TOWN LOTS

In San Francisco, Sonoma, Pueblo de San Jose, Monterey, Benicia, Saucelito, Sutterville and Stockton, for sale.

250

THOUSAND dollars, to loan on Bond and Mortgage.

Consignments of Merchandise &c., attended to and liberal advances made thereon, by

ROACH & WOODWORTH,
PARKER HOUSE.

SAN FRANCISCO, MARCH 3, 1849.

San Francisco was an instant city: by the fall of 1849 it had not only saloons and restaurants but even a lemonade factory. But, despite what people in the East thought, the streets were not paved with gold. In fact, they were not paved at all, and in the dank winter of 1849 it rained for 71 straight days, turning the streets into muddy troughs. Some of the mud holes and puddles were so deep that people fell in and drowned.

Montgomery Street, a sign said, was impassable. "Not even jackassable."

Most of the people who came to San Francisco that golden year came by sea. When they were old and gray they were hailed as pioneers, but the ship passengers were regarded with suspicion by Californians of the day.

"Speaking broadly," Bancroft wrote, "with all due regard to exceptions . . . [sea travelers] embraced more of the abnormal and ephemeral and a great deal of the criminal and vicious in early California life.

(above) The curse of the mines: guns, gambling, and dice. This classic pistol had a revolving barrel and a short wooden stock.

(below) "Gambling, drinking, and houses of ill fame are the chief amusements of this country," wrote Lucius Fairchild in 1850. Wallace Baker (left) posed for this portrait with a winning poker hand.

"They might build cities and organize society but there were among them those who made the cities hotbeds of vice and corruption and who converted the social fabric into a body at the sight of which the world stood wrapped in apprehension."

The crime rate in the gold rush period was horrendous. In the 1850s, for example, there were 44 murders in 15 months in Los Angeles County, which had a population of only 8,500. In the whole state, the official murder count was 500 to 550 a year, at a time when there were 300,000 non-Indians in California, a rate 17 times that of modern California.

In a book called *The Land of Gold,* published in 1854, it was estimated that "since the opening of the mines" there were 4,200 murders, 1,400 suicides, and "ten thousand miserable deaths." How many Indians were killed, no one knows, but the native population declined from 150,000 in 1848 to about 36,000 in just 12 years.

History is kinder to those who came to California by land. Their road was harder, and Bancroft, among others, says that they were better people. They came, he said, "from small towns, villages of the interior . . . honest, industrious, and self reliant."

(above) Sharpshooters in uniform from the 1856 San Francisco Committee of Vigilance. Citizens often took the law into their own hands in California—the 1856 San Francisco vigilantes even banged a city supervisor.

(above) A single man against the odds: A.D.O. Browere's Lone Prospector, *painted in 1853, shows the gold seeker in his glory— a mule, a pan, a shovel, a pistol, a rifle, a hope.*

No one who lacked these qualities could make it, apparently, because to come by land one had to walk.

The overland pioneers, who hailed mostly from the South and Midwest, also thought of themselves as high-minded men and women.

John Hittell, who delivered the oration on California's Admission Day celebration in 1869, saw the vast numbers of travelers to California who were poised on the Missouri River frontier in the spring of 1849 as "the flower of the west, nearly all young, active, healthy, many well educated, all full of hope and enthusiasm."

Hittell went on: "In our ignorance of the nature of auriferous deposits, we expected . . . to strike places where we should dig up 200 to 300 pounds of gold a day without difficulty. . . . In visions by day and in dreams by night we saw ourselves in possession of treasures more splendid than those that dazzled the eyes of Aladdin. We compared ourselves to the Argonauts, to the army of Alexander starting to conquer Persia, to the Crusaders."

Another said the travelers west—the pioneers—began a journey across a land of which they knew little to work a trade of which they knew nothing.

In 1849, the first big year of the gold rush, at least 25,000 people, mostly men, went overland to California. Some put the number as high as 45,000. Nearly all of them traveled by wagon, though some took pack animals. They carried supplies for a year, along with all kinds of possessions—pianos, mining tools, ovens, bottles, furniture. To survive the journey, most of it was thrown away on the trail.

Legend has it the wagon trains were attacked on their westward journey by fierce plains Indians, but the real enemy during the gold rush was disease. Thousands of pioneers died on the way, most from a particularly vicious form of cholera that followed them across the Missouri River and into the West.

Many of the 49ers kept diaries, all of which mention the curse of cholera, a disease that had no known cure, one that would strike a man in the morning and kill him by nightfall. One member of a company of Midwesterners heading for California, D. Jagger, stopped at a funeral on the plains only a couple of weeks into the wagon trip west.

"We came upon a collection of people around an open grave, attending the funeral of a distinguished Freemason. . . . At the close they sang the well-known dirge, 'Thou art gone to the grave . . .' [and] the lone hills

(above) The wooden ox yoke was used to pull wagons overland to reach California.

(below) Staring ahead at the unknown California: a prospector and a miner in the Sierra. This is called Mountain Jack and a Wandering Miner, *by E. Hall Martin.*

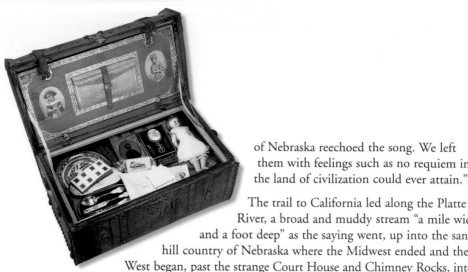

(above) Immigrants to California brought their own treasures to the land of gold. This one is a steamer trunk brought around Cape Horn, filled with dolls and pictures of loved ones.

(below) The gold rush drew painters as well as miners. This portrait of the gold camp of Forbestown was done by A. Findlay.

of Nebraska reechoed the song. We left them with feelings such as no requiem in the land of civilization could ever attain."

The trail to California led along the Platte River, a broad and muddy stream "a mile wide and a foot deep" as the saying went, up into the sand-hill country of Nebraska where the Midwest ended and the West began, past the strange Court House and Chimney Rocks, into Wyoming.

The travelers wrote their names on the cliffs, and on the noted Independence Rock on the banks of the Sweetwater River in Wyoming, where their words and the ruts of their wagon wheels can be seen even today.

They crossed the continental divide at South Pass—and unknowingly walked right by land rich in gold and silver.

They went down the western slope, past the beautiful Wind River Mountains, stopping at Fort Hall, the last building for a thousand miles, then moved west and west, following the setting sun like the old man in the Steinbeck story "The Leader of the People," westering and westering.

They moved down the brown and dreary Humboldt River in Nevada— "three hundred miles and a cloud of dust," they wrote.

Some followed the advice of Peter Lassen, another California pioneer, and took a shortcut that led only to fierce desert. They cursed his name.

Another party coming from the Southwest entered an even harsher desert and nearly lost everything. When they left that poisoned land, one looked back and called out, "Good-bye, Death Valley!" giving the place its name.

The rest crossed the Sierra just before the snows and caught sight of California. "I thought it was the grandest scene I ever saw," said P. F. Castleman.

Weeks later, Castleman arrived at the actual gold diggings. "We travelled until after dark when we came to the encampments at Long's Bar. It reminded me of a city when illuminated to its height. As we passed through a ravine both sides seemed literally covered with tents. All had large fires near them which made it seem almost as light as day and all seemed to be life and glee as a thousand voices seemed to mingle together, some talking, laughing and singing, these with roaring of waters as they dashed over the stony beds and against the rockbound shores of this river seemed wild and romantic to my ears. . . ."

(above) Using water brought in on long flumes, miners washed away soil to get down to bedrock—and perhaps gold. This operation was in El Dorado County.

(below) Grave marker of James McDowell, dated May 1849.

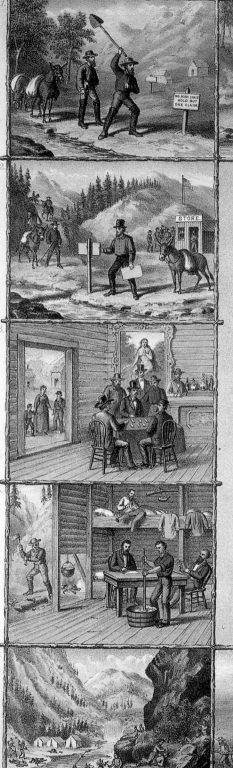

A MAN SPAKE THESE WORDS, AND SAID: I am a miner who wandered from "Away Down East," and came to sojourn in a strange land and "See the Elephant." And behold I saw him, and bear witness that, from the key of his trunk to the end of his tail, his whole body has passed before me; and I followed him until his huge feet stood still before a clapboard shanty; then, with his trunk extended, he pointed to a candle-card tacked upon a shingle, as though he would say **"READ!"** and I read the

PIONEERS' TEN COMMANDMENTS.

I.

Thou shalt have no other claim than one.

II.

Thou shalt not make unto thyself any false claim, nor any likeness to a mean man by jumping one. Whatever thou findest, on the top above, or on the rock beneath, or in a crevice underneath the rock, or I will visit the miners around to invite them on my side; and when they decide against thee, thou shalt take thy pick, thy pan, thy shovel, and thy blankets, with all that thou hast, and go prospecting to seek good diggings; but thou shalt find none. Then, when thou hast returned, in sorrow shalt thou find that thine old claim is worked out, and yet no pile made thee to hide in the ground or in an old boot beneath thy bunk, or in buckskin or bottle underneath thy cabin; but has paid all that was in thy purse away, worn out thy boots and thy garments, so that there is nothing good about them but the pockets, and thy patience is likened unto thy garments; and at last thou shalt hire thy body out to make thy board and save thy bacon.

III.

Thou shalt not go prospecting before thy claim gives out. Neither shalt thou take thy money, nor thy gold dust, nor thy good name, to the gaming table in vain; for monte, twenty-one, roulette, faro, lansquenet and poker will prove to thee that the more thou puttest down the less thou shalt take up; and when thou thinkest of thy wife and children, thou shalt not hold thyself guiltless, but—insane.

IV.

Thou shalt not remember what thy friends do at home on the Sabbath day, lest the remembrance may not compare favorably with what thou doest here. Six days thou mayest dig or pick all that thy body can stand under, but the other day is Sunday; yet thou washest all thy dirty shirts, darnest all thy stockings, tap thy boots, mend thy clothing, chop thy whole week's fire-wood, make up and bake thy bread and boil thy pork and beans that thou wait not when thou returnest from thy long-tom weary. For in six days' labor only thou canst not work enough to wear out thy body in two years; but if thou workest hard on Sunday also, thou canst do it in six months; and thou and thy son and thy daughter, thy male and thy female friend, thy morals and thy conscience be none the less better for it, but reproach thee shouldst thou ever return to thy mother's fireside; and thou strive to justify thyself because the trader and

the blacksmith, the carpe Jews and Buccaneers def the Sabbath day, nor wis of youth and and home ma

Thou shalt not think canst make it fastest, than ridden rough-shod over th amples, that thou mayest thee when thou art left blessing and thy mothers'

Thou shalt not kill th though thou shalt make e with. Neither shalt thou for by keeping cool thou c Neither shalt thou destre "slewed," nor "high," i nor "three sheets in the "brandy slings," "gin co toddies" nor "egg nogs." juleps" nor "sherry cobble a bottle the raw material, while thou art swallowing off thy back, thou art burr and if thou couldst see the home comforts already lyi feel a choking in thy thro crooked walking and hic broiling in the sun, of pro shafts and ditches from wh ing rat, thou wilt feel dis thy servant a dog that he say, farewell old bottle; I and thou, slings, cock-tail toddies, sangarees and jule brance shames me; hencef headaches, tremblings, he unholy catalogue of evils w smiles and my children's m reward me for having the "No! I wish thee an eter

Thou shalt not grow c before thou hast made t "struck a lead" nor found a "pocket," lest in going and go to work ashamed right; for thou knowest b a lead and fifty dollars a d and then go home with happy.

COPYRIGHTED 1887 BY W.P. BENNETT, GOLD HILL, NEVADA.

SCENES WHEN CRO

THE MINER'S PIONEER T

e merchant, the tailors,
vilization by keeping not
ot rest, such as memory

l thy gold, nor how thou
lt enjoy it after thou hast
parents' precepts and er,
g to reproach and sting
land where thy father's
nt thee.

working in the rain, even
y physic and attendance
ghbor's body in a duel,
s life and thy conscience.
y getting "*tight*," nor
," nor "*half-seas over*,"
lrinking smoothly down
whisky punches," "*rum*
shalt thou suck "*mint*
a straw, nor gurgle from
leat from a decanter, for
purse and thy coat from
t from off thy stomach;
lands, and gold dust, and
huge pile—thou shouldst
t to that thou add'st thy
lodging in the gutter, or
alf full of water, and of
t emerged like a drown-
thyself, and inquire, "*Is
hings?*" Verily, I will
gurgling lips no more;
smashes, cobblers, nogs,
farewell. Thy remem-
at thy acquaintance; and
, blue-devils, and all the
n thy train. My wife's
laugh shall charm and
ess and courage to say:
!"

nor think of going home
because thou hast not
ce," nor sunk a hole upon
eave four dollars a day
s a day, and serve thee
ere thou mightest strike
o thy manly self-respect,
ake thyself and others

VIII.

Thou shalt not steal a pick, or a pan, or a shovel, from thy
fellow miner, nor take away his tools without his leave; nor
borrow those he cannot spare; nor return them broken; nor
trouble him to fetch them back again; nor talk with him while
his water rent is running on; nor remove his stake to enlarge
thy claim; nor undermine his claim in following a lead; nor pan
out gold from his riffle-box; nor wash the tailings from the
mouth of his sluices. Neither shalt thou pick out specimens
from the company's pan to put in thy mouth or in thy purse;
nor cheat thy partner of his share; nor steal from thy cabin-
mate his gold dust to add to thine, for he will be sure to dis-
cover what thou hast done, and will straightway call his fellow
miners together, and if the law hinder them not they will hang
thee, or give thee fifty lashes, or shave thy head and brand
thee like a horse thief with "R" upon thy cheek, to be known
and of all men Californians in particular.

IX.

Thou shalt not tell any false tales about "*good diggings
in the mountains*" to thy neighbor, that thou mayest benefit a
friend who hath mules, and provisions, and tools, and blankets
he cannot sell; lest in deceiving thy neighbor when he returns
through the snow, with naught but his riffle, he present thee
with the contents thereof, and like a dog thou shalt fall down
and die.

X.

Thou shalt not commit unsuitable matrimony, nor covet
"*single blessedness*," nor forget absent maidens, nor neglect thy
first love; but thou shalt consider how faithfully and patiently she
waiteth thy return; yea, and covereth each epistle that thou
sendeth with kisses of kindly welcome until she hath thyself.
Neither shalt thou covet thy neighbor's wife, nor trifle with
the affections of his daughter; yet, if thy heart be free, and thou
love and covet each other, thou shalt "*pop the question*" like
a man, lest another more manly than thou art should step in before thee, and thou lovest her in vain, and, in the anguish of thy
heart's disappointment, thou shalt quote the language of the
great, and say, "*sich is life;*" and thy future lot be that of a
poor, lonely, despised and comfortless bachelor.

A new commandment give I unto you. If thou hast a wife
and little ones, that thou lovest dearer than thy life, that thou
keep them continually before you to cheer and urge thee
onward until thou canst say, "*I have enough; God bless them;
I will return.*" Then as thou journiest towards thy much loved
home, with open arms, shall they come forth to welcome thee,
and falling on thy neck, weep tears of unutterable joy that thou
art come; then in the fullness of thy heart's gratitude thou shalt
kneel before thy Heavenly Father together, to thank Him for
thy safe return. Amen. So mote it be.

KURZ & ALLISON'S ART STUDIO, CHICAGO, U.S.A.

(above) The assay office
made the key judgment—
was the ore worth a
fortune or nothing? This
office belonged to James
J. Ott, a Swiss business-
man who settled in
Nevada City.

(previous pages)
The Miner's Pioneer Ten
Commandments of 1849.

They founded dozens of towns and camps
with names like Timbuctoo, You Bet, Coyote
Hill and North San Juan, Negro Bar, Keno and
Jimmy Brown, Brass Wire, Whiskey, Brandy, Jackass,
Lizard and Virgin Flats, Poorman's Creek—where a million dol-
lars in gold was found—and Gold Flat, Eureka, Sailor Flat, Rich Flat,
Boston Ravine and Auburn Ravine, Sucker Flat, Excelsior, Poker Flat,
and Whisky Diggins.

"We had come to dig for gold," said John Hittell, who had compared his
companions to Argonauts and Crusaders, "and nearly all of us who came
by land went to mining. Though we did not make so much as we had
hoped, we found the placers wonderfully rich. It was no uncommon
event for a man alone to take out five hundred dollars in a day or two or
three, if working together, to divide the dust at the end of the week by
measuring it with tin cups, but we were never satisfied."

There were big stories. At Rich Bar on the Feather prospectors found
$2,900 in two pounds of dirt, $15,000 in two days.

There was gold in the north, on the beach at Crescent City, on the rivers
near Weaverville, a now-quiet town that had 4,000 people in 1850.
Mokelumne Hill was a vertical mountain of gold; Table Mountain in the
southern Mother Lode was "immensely rich with coarse gold of the
highest quality."

But after the tall tales of bonanzas and fortunes to be made, there really was very little gold that could be picked up using the primitive methods of the day.

Hittell and his companions went on deeper in the mountains, where they found diggings "that would make millionaires of us." Then they ran out of food. They could not eat gold, of course. "We had to live for days on grass and acorns picked from the holes in the trees where they had been placed by woodpeckers."

For months they "slept in no shelter and saw no house." Nor did they get the gold they wanted.

The miners worked the rivers, dug them up, ran ditches, worked in the pitiless summer sun and the damp winter. Most of them found nothing.

(above) Adams and Company Bank Certificate, dated February 6, 1854.

(below) Lucky miners could find nuggets of solid gold in the rivers and streams. Some of them made the nuggets into jewelry, like these stick-pins worn by two unknown miners.

(above) A miner's life was not all work. Here is a Sunday diversion—drinking, talking, and dancing. There were few women, so sometimes young men dressed up in faux skirts and danced with other men.

But it was an adventure.

Mark Twain said it best in *Roughing It*: "It was a splendid population—for all the slow, sleepy, sluggish-brained sloths staid at home—you never find that sort of people among pioneers—you cannot build pioneers out of that sort of material.

"It was that sort of population that gave to California a name for getting up an astounding enterprise and rushing them through with a magnificent dash and daring and a recklessness of cost or consequences which she bears to this day—and when she projects a new surprise, the grave world smiles as usual and says, 'Well, that is California all over.'

"But they were rough in those mines! They fairly reveled in gold, whisky, fights, fandangoes, and were unspeakably happy."

That was the myth. In the fall of 1850, cholera was supposed to have carried off 15 percent of the population of Sacramento, 10 percent of San Jose, and 5 percent of San Francisco.

And the miners eventually gave up.

"They drifted along," Bancroft said, "in semi-beggery from snow clad ranges to burning plain, brave and hardy, gay and careless, till lonely age crept up to confine them to some rural hamlet, emblematic of their shattered hopes. . . ."

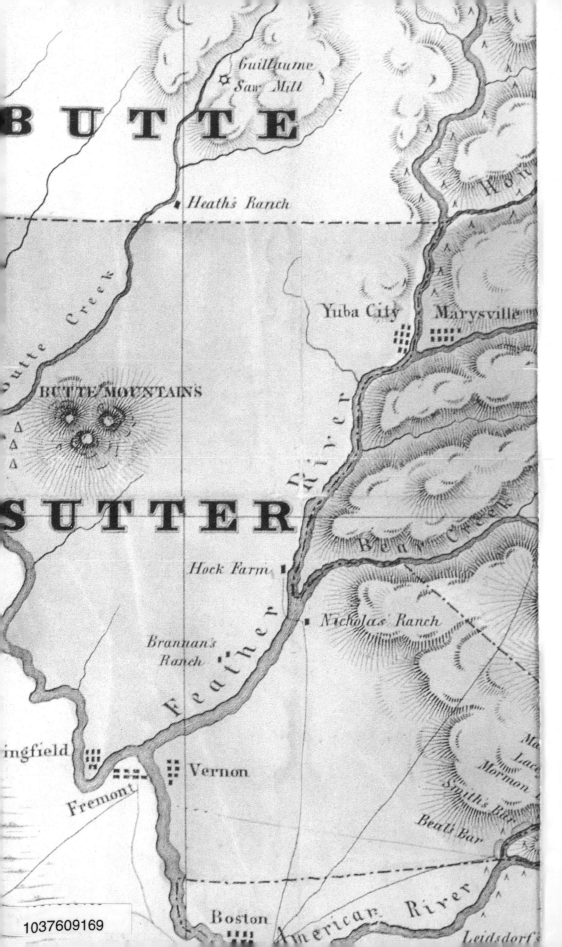

Mining, as it turned out, really didn't pay. "One has merely to divide the total annual production by the number of workers to find their earnings were far below the current wages," Bancroft said.

In 1852, he explained, the average earning was $600 a year and the average rate $1 a day for miners, compared to the average wage in non-mining activities of $4 to $5 a day.

Those whose lives were first touched by gold fared worst.

John Sutter lost his land in the gold rush and spent his last days in the East, far from California.

Californians had honored Sutter by naming a county after him, but the gold seekers had taken everything else. He even lost the sawmill where gold was found, the little empire he had built in the Sacramento valley, and it ruined him, he said. Bancroft interviewed Sutter and listened to his sad story. He concluded that Sutter was impecunious, "his own worst enemy," not quick enough to protect his interest, not shrewd enough to see the main chance. Sutter had had a golden opportunity but, Bancroft noted, "was not man enough to grasp and master his good fortune."

James Marshall, the man who first discovered gold, was equally unlucky. He was an odd duck, anyway, and the new miners found it difficult to deal with him. He tried to defend the Indians and was run out of his

(above) Issac Wallace Baker, a pioneer photographer, posed for his own picture in front of another photographer's shop, Batchelder's Daguerrian Saloon.

(lower opposite) The mine workers wore their tools as soldiers wore their arms. This miner, in his Sunday best, posed for a picture with his shovel.

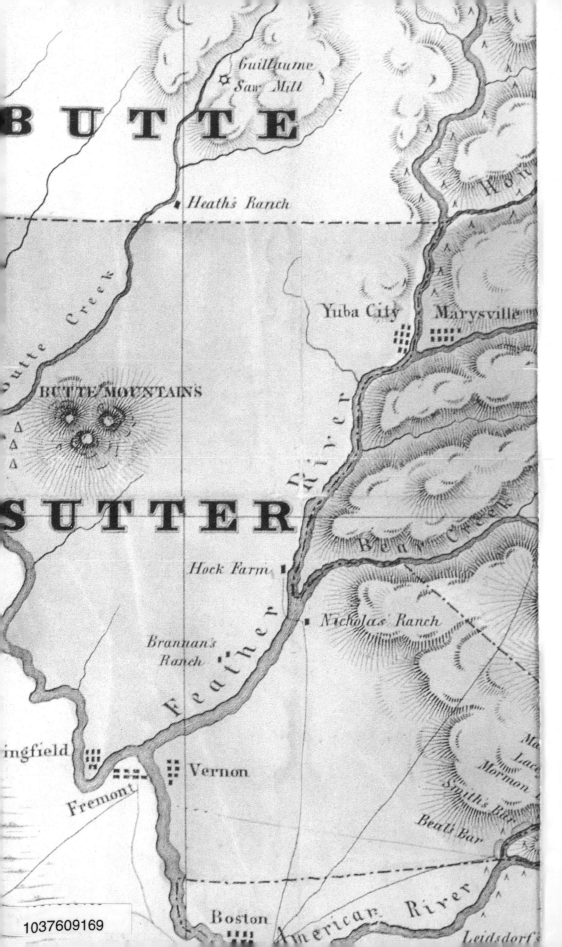

1037609169

(above)
These row houses in South Park marked San Francisco's first neighborhood for well-off people. The park is still there, but the houses have been replaced by modern businesses and dwellings.

own mining camp. Marshall never found gold again; he drank, he complained of his lot, he died alone and broke.

So what was left? California was left. Sallie Hestor, who made the five-month trip with her family from St. Joseph, Missouri, in 1849, described her family's outlook: "We are strangers in a strange land—what will our future be?"

They moved to San Jose, far from the gold fields. Her father became the first district attorney of Santa Clara County and later a judge. Sallie married and lived the rest of her long life in California.

The heroic pioneers ended up on monuments, Sutter and Marshall among them. The miners ended up in cemeteries—you can still see their graves in the little towns of the Mother Lode. The era died. "Other years have been repeated," said Bancroft of the famous year of gold, 1849. "This one, never."

A CHRONOLOGY OF THE GOL

- 1848 -

January 24
James Marshall discovers the first gold nugget a

March 1
News of gold discovery reaches San Fran

December 5
President Polk reports gold discovery to C

- 1849 -

December 20
San Francisco harbormaster records 782 ship arrivals

- 1850 -

California becomes 31st state in the Compromi

- 1854 -

Sacramento chosen as state capital

- 1856 -

First railroad in California built,
22 miles between Sacramento and Folso

Mining, as it turned out, really didn't pay. "One has merely to divide the total annual production by the number of workers to find their earnings were far below the current wages," Bancroft said.

In 1852, he explained, the average earning was $600 a year and the average rate $1 a day for miners, compared to the average wage in non-mining activities of $4 to $5 a day.

Those whose lives were first touched by gold fared worst.

John Sutter lost his land in the gold rush and spent his last days in the East, far from California.

Californians had honored Sutter by naming a county after him, but the gold seekers had taken everything else. He even lost the sawmill where gold was found, the little empire he had built in the Sacramento valley, and it ruined him, he said. Bancroft interviewed Sutter and listened to his sad story. He concluded that Sutter was impecunious, "his own worst enemy," not quick enough to protect his interest, not shrewd enough to see the main chance. Sutter had had a golden opportunity but, Bancroft noted, "was not man enough to grasp and master his good fortune."

James Marshall, the man who first discovered gold, was equally unlucky. He was an odd duck, anyway, and the new miners found it difficult to deal with him. He tried to defend the Indians and was run out of his

(above) Issac Wallace Baker, a pioneer photographer, posed for his own picture in front of another photographer's shop, Batchelder's Daguerrian Saloon.

(lower opposite) The mine workers wore their tools as soldiers wore their arms. This miner, in his Sunday best, posed for a picture with his shovel.

(above)
These row houses in South Park marked San Francisco's first neigh-borhood for well-off people. The park is still there, but the houses have been replaced by modern businesses and dwellings.

own mining camp. Marshall never found gold again; he drank, he complained of his lot, he died alone and broke.

So what was left? California was left. Sallie Hestor, who made the five-month trip with her family from St. Joseph, Missouri, in 1849, described her family's outlook: "We are strangers in a strange land—what will our future be?"

They moved to San Jose, far from the gold fields. Her father became the first district attorney of Santa Clara County and later a judge. Sallie married and lived the rest of her long life in California.

The heroic pioneers ended up on monuments, Sutter and Marshall among them. The miners ended up in cemeteries—you can still see their graves in the little towns of the Mother Lode. The era died. "Other years have been repeated," said Bancroft of the famous year of gold, 1849. "This one, never."

A Chronology of the Gold Rush

- 1848 -

January 24
James Marshall discovers the first gold nugget at Sutter's Mill

March 1
News of gold discovery reaches San Francisco

December 5
President Polk reports gold discovery to Congress

- 1849 -

December 20
San Francisco harbormaster records 782 ship arrivals since March 26

- 1850 -

California becomes 31st state in the Compromise of 1850

- 1854 -

Sacramento chosen as state capital

- 1856 -

First railroad in California built,
22 miles between Sacramento and Folsom

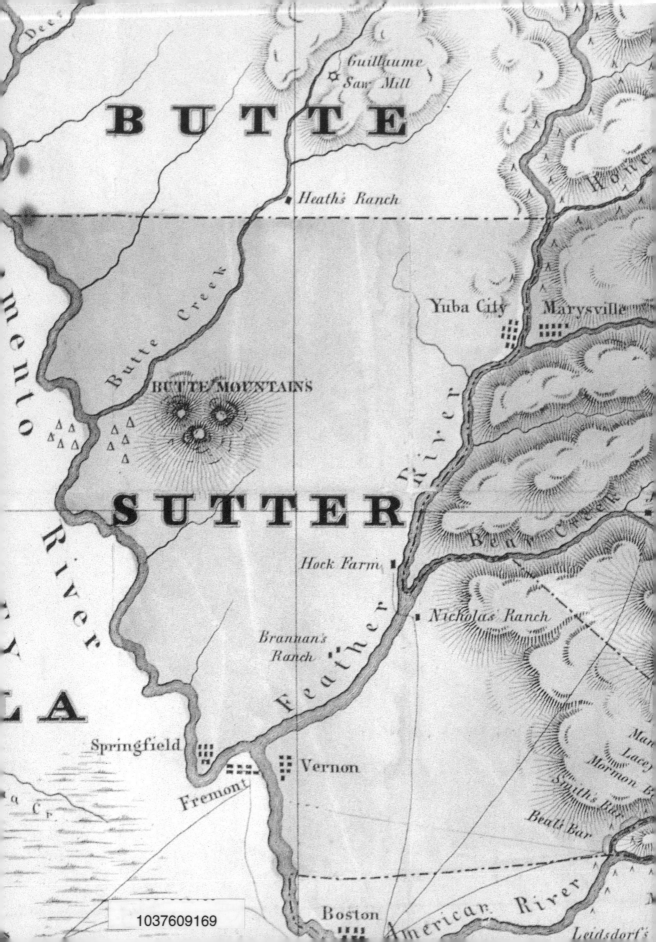

BUTTE

Guillaume Saw Mill

Heath's Ranch

Butte Creek

BUTTE MOUNTAINS

Yuba City   Marysville

SUTTER

Hock Farm

Nicholas Ranch

Bear Creek

Feather River

Brannan's Ranch

Springfield

Vernon

Fremont

Lacey
Mormon B.
Smith's Bar
Beal's Bar

Boston

American River

Leidsdorff's

1037609169